U0231956

KITCHEN Ⅱ

新空间·厨房 Ⅱ

编者：新空间编辑组

辽宁科学技术出版社

KITCHEN Ⅱ

新空间・厨房 Ⅱ

编者：新空间编辑组

辽宁科学技术出版社

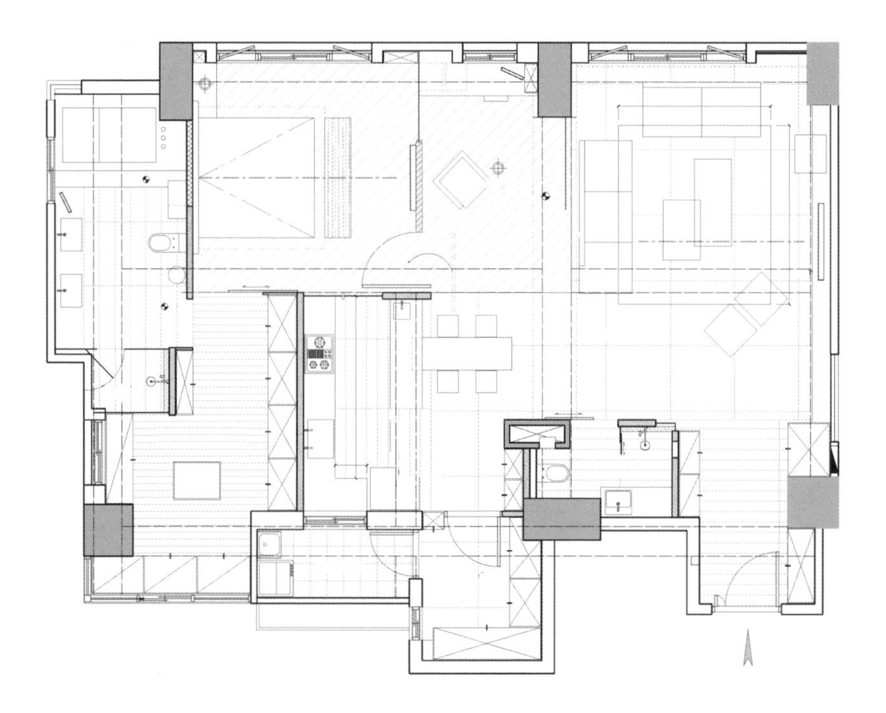

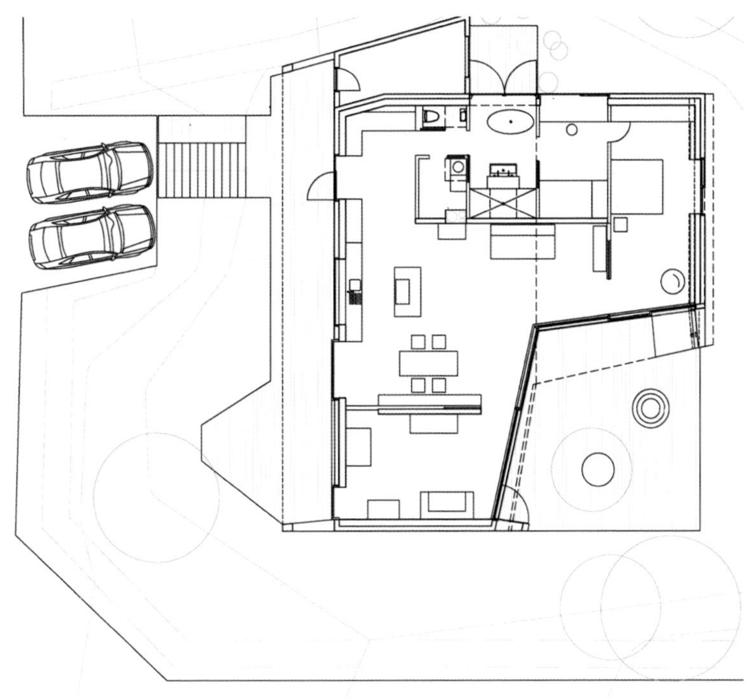

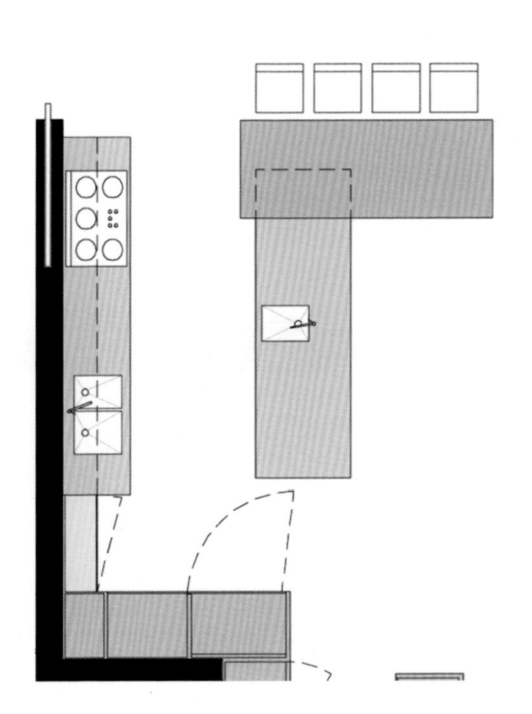

Omelette
tablets 8mg
10 tablets
HirstDamien

Wallpaper*

ROTHKO

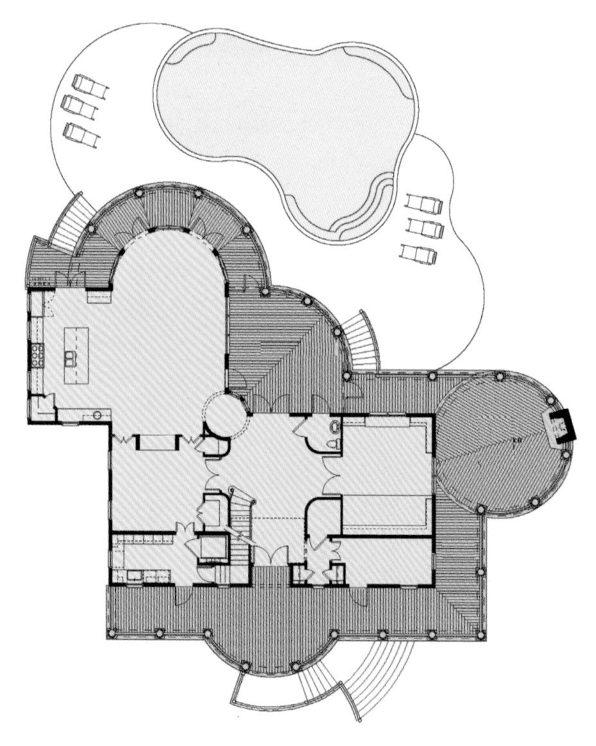

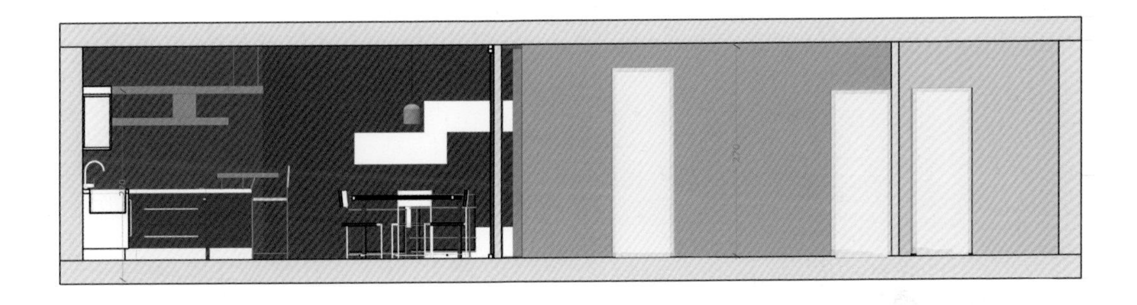

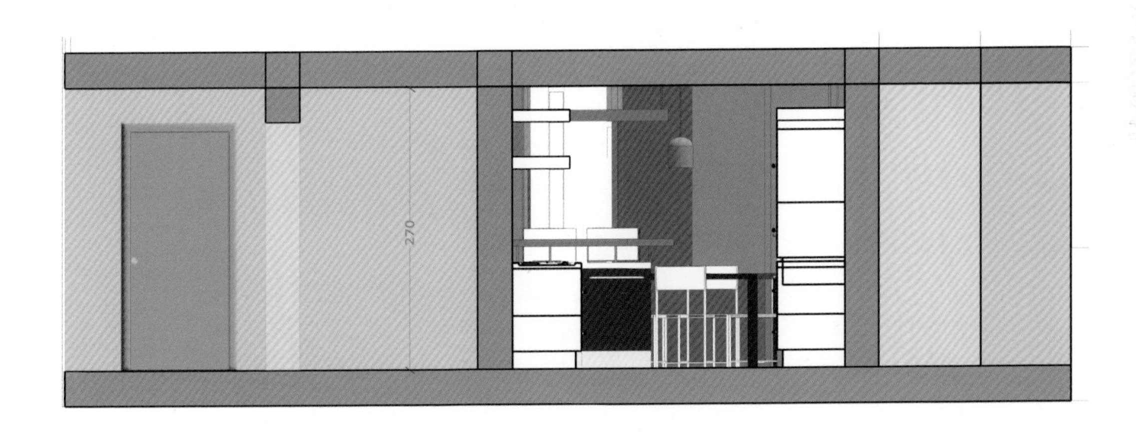
270

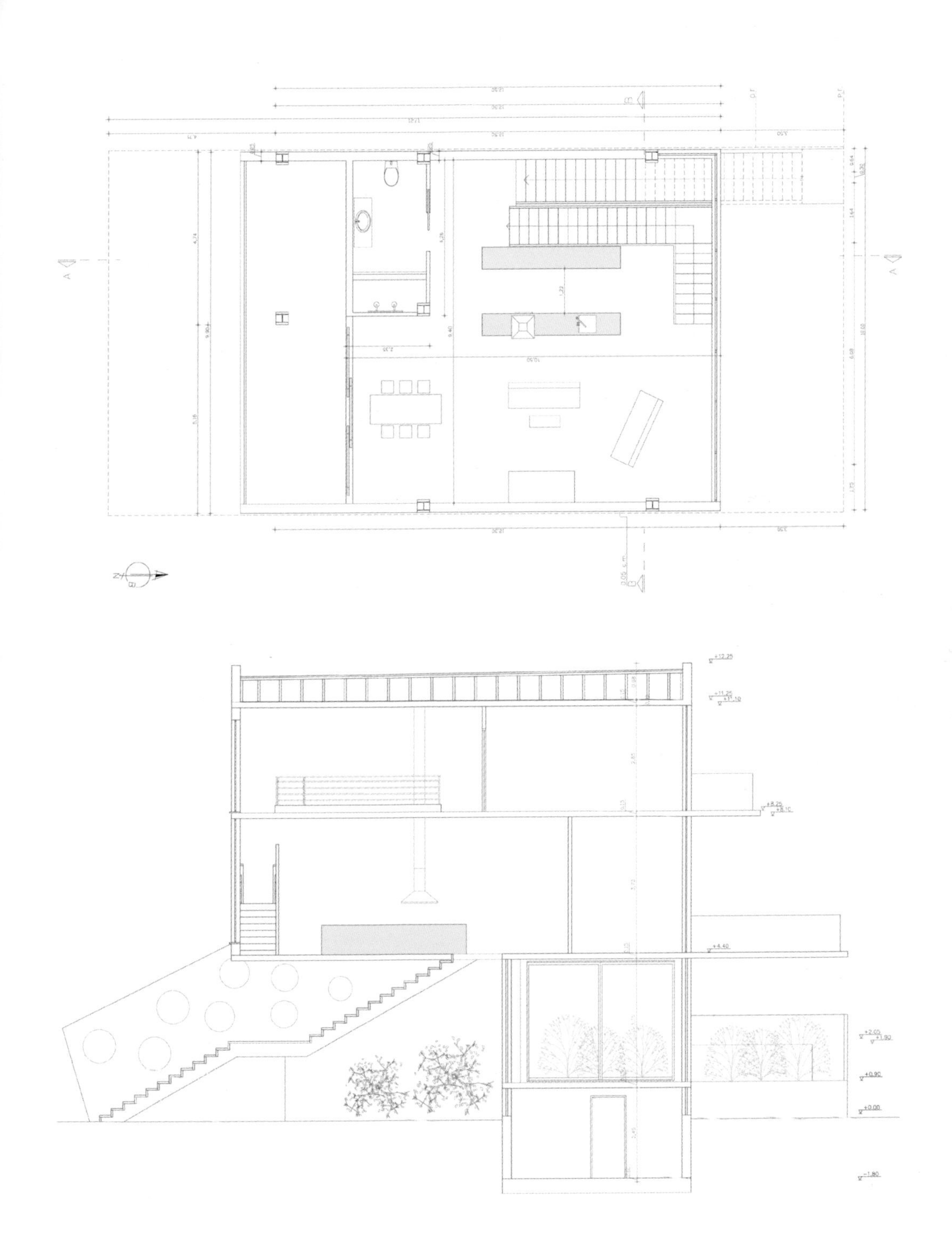

061

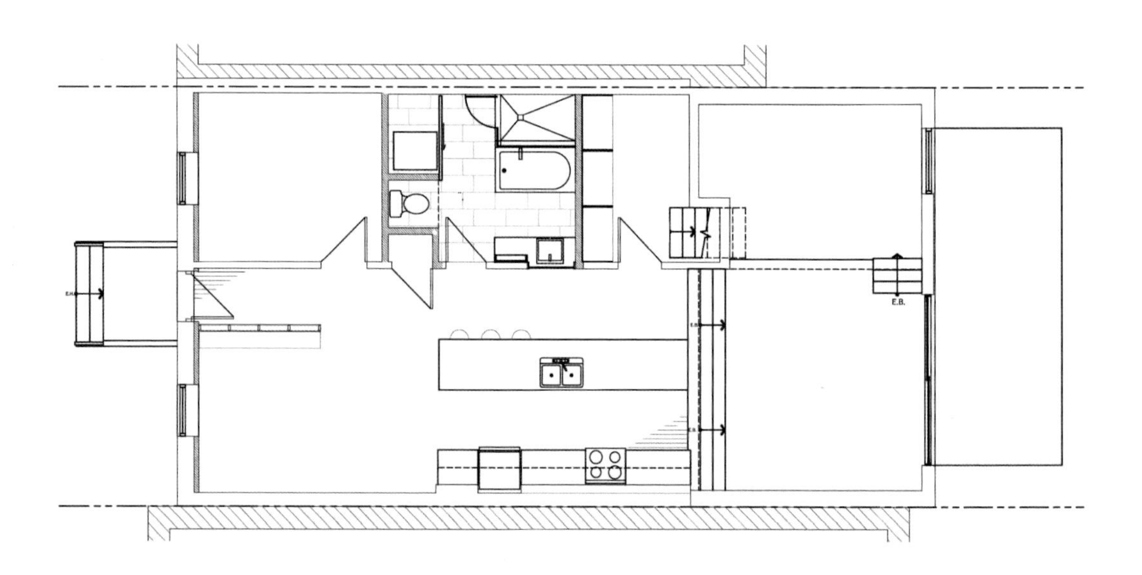
E.B.

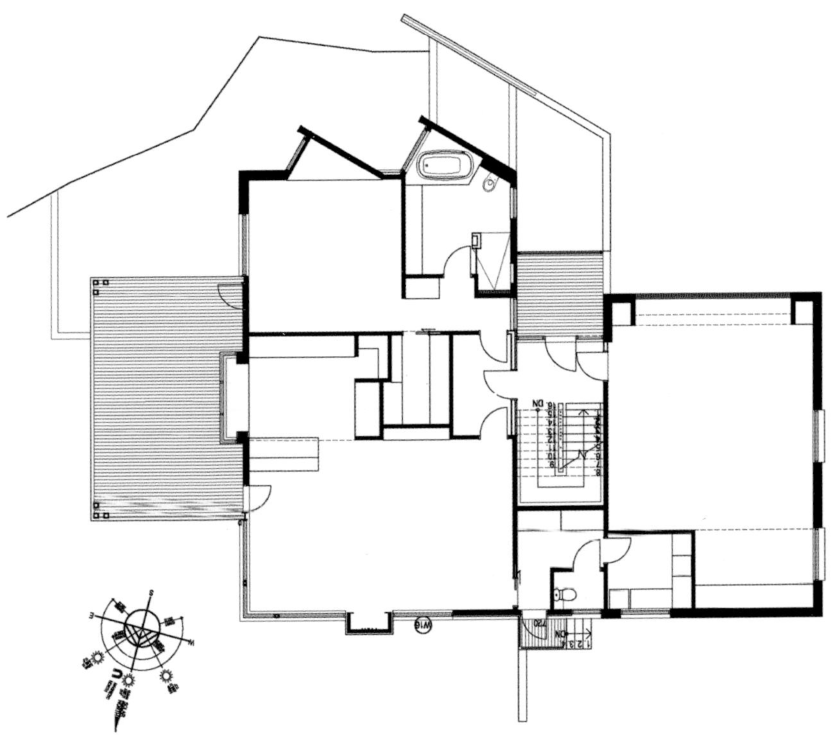

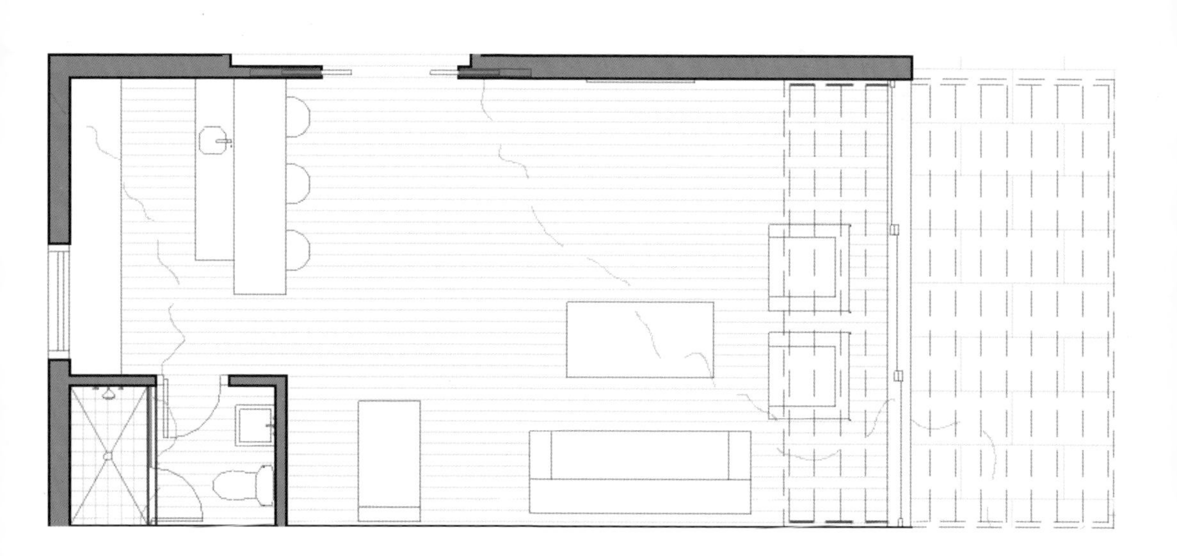

les radis croquent
les tomates rougissent !!!

les carottes
les tomates
rougissent !!!

BOSCH

Tom Clancy
THE TOMMYKNOCKERS
THE TRAVELER
TALISMAN
BLACK HEART
THE LIMERICK
CLIVE CUSSLER

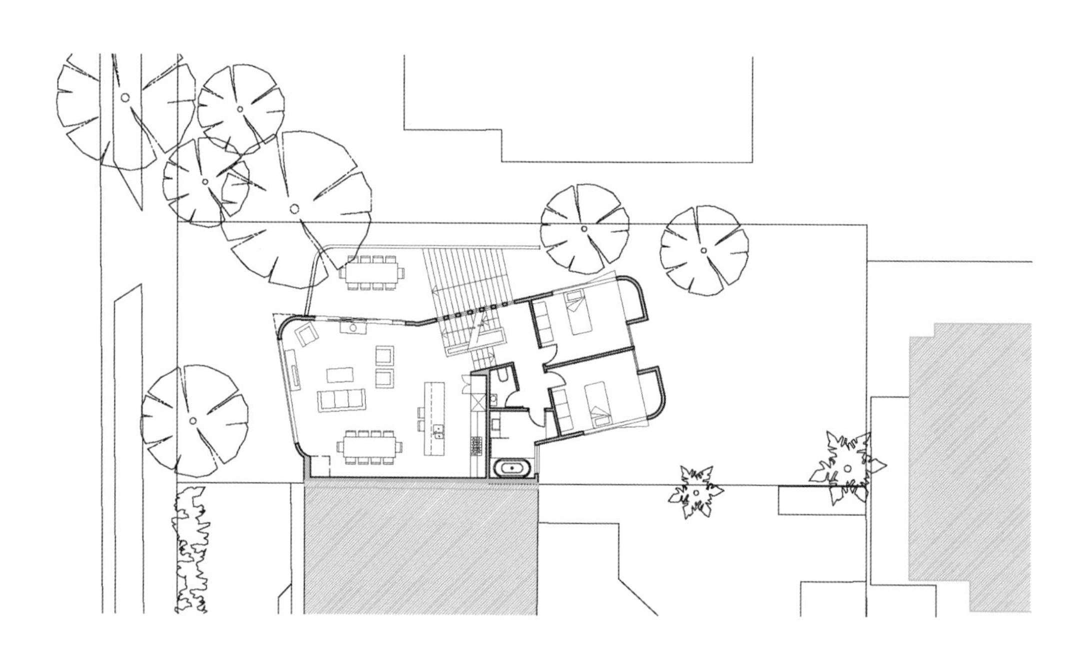

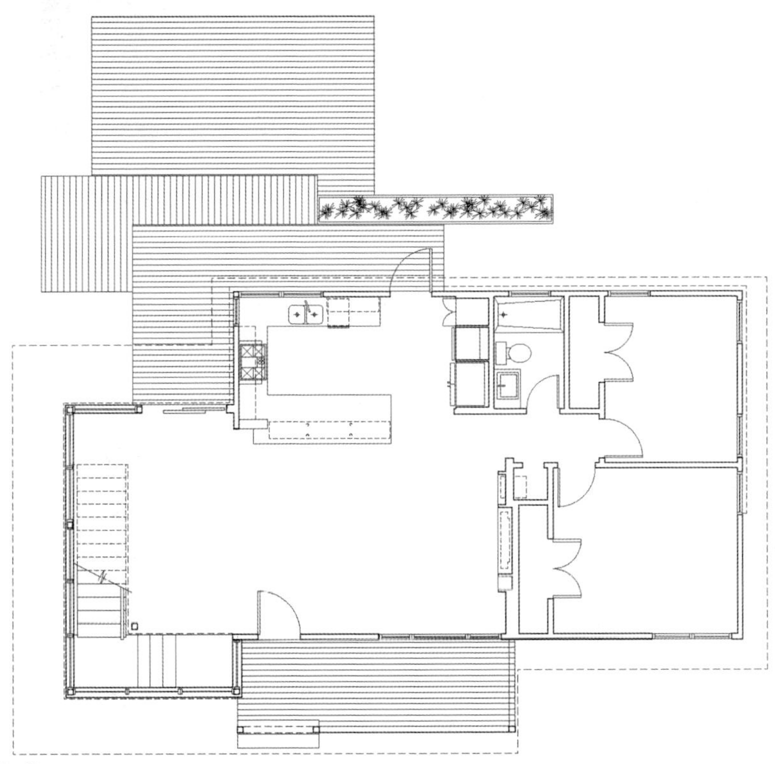

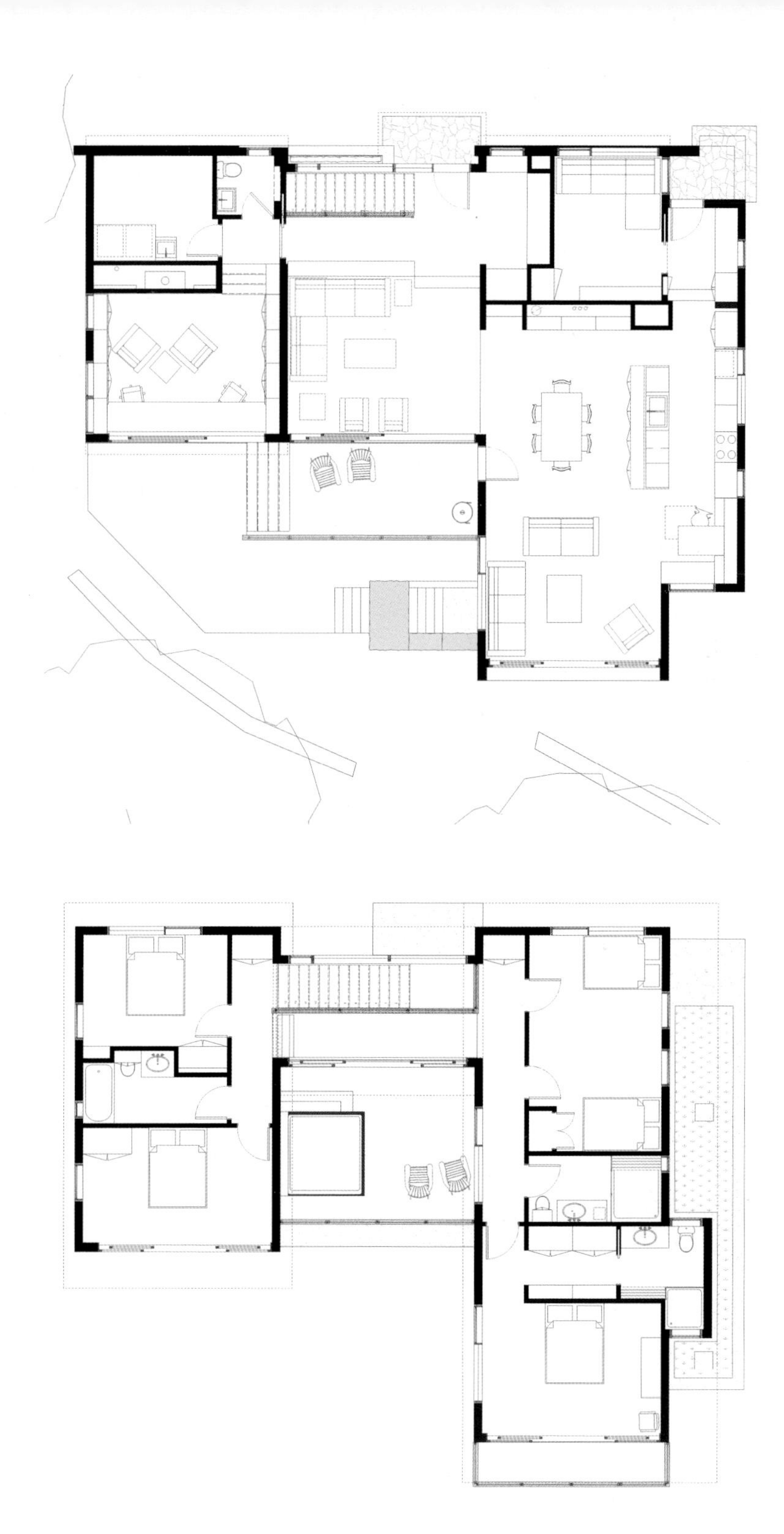

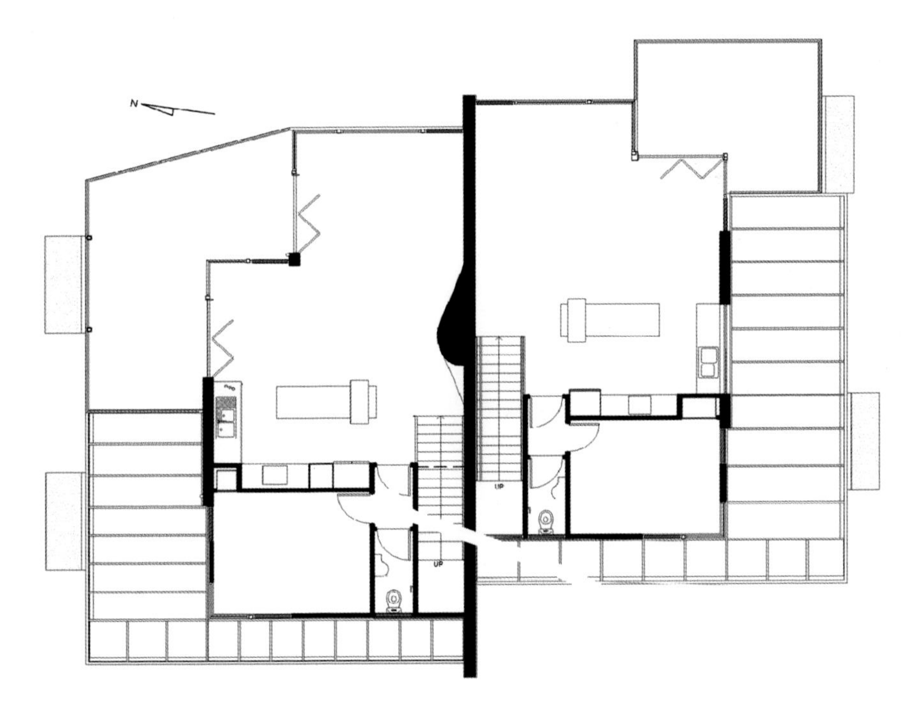
N
UP
UP

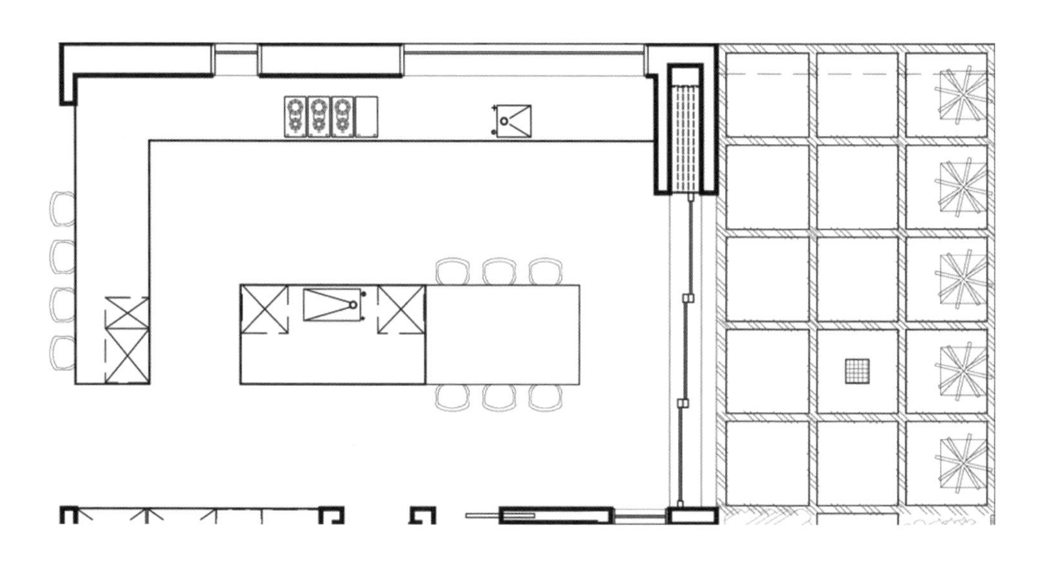

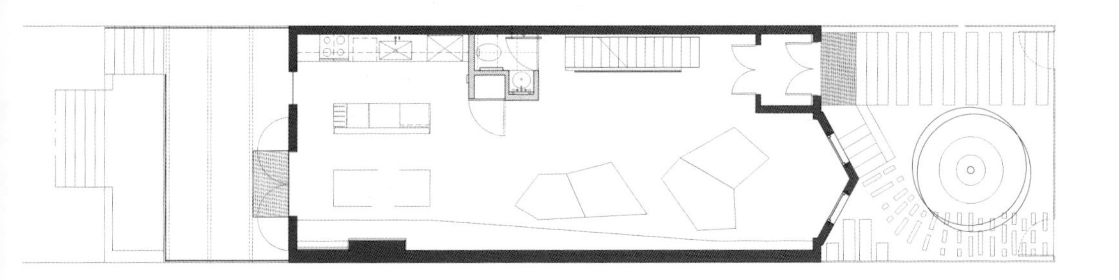

SPOON

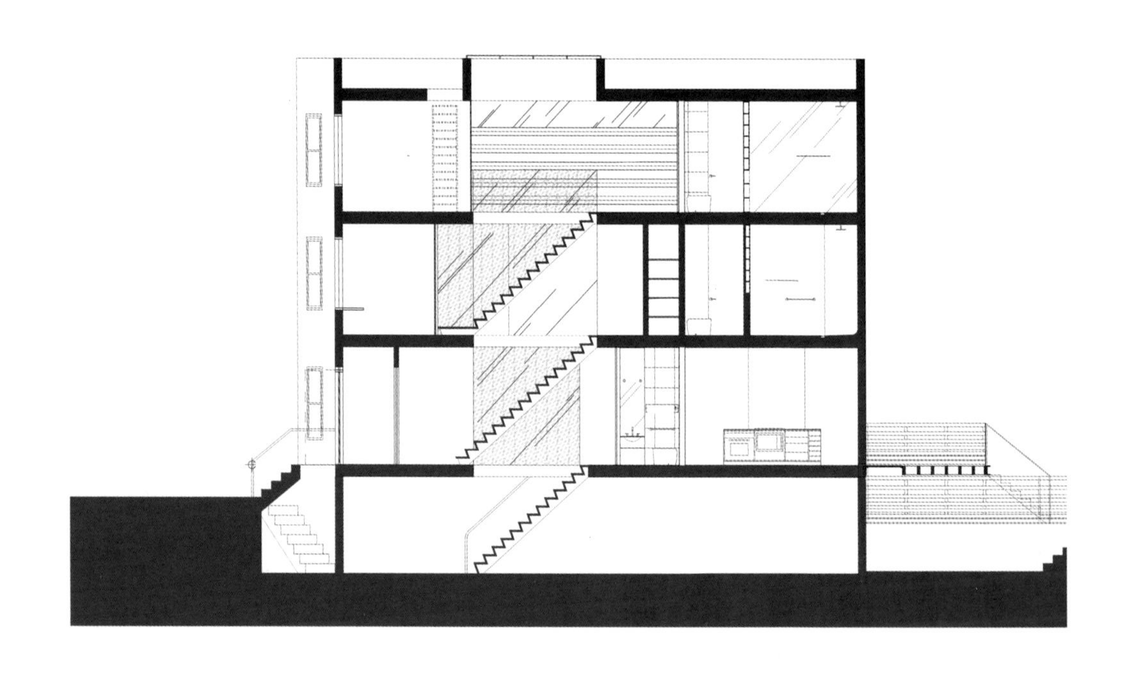

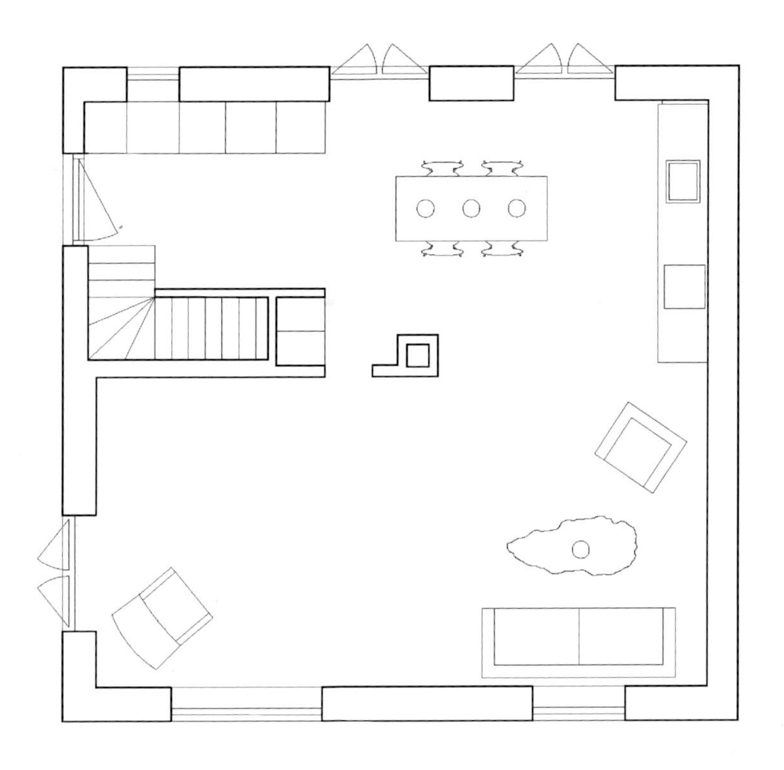

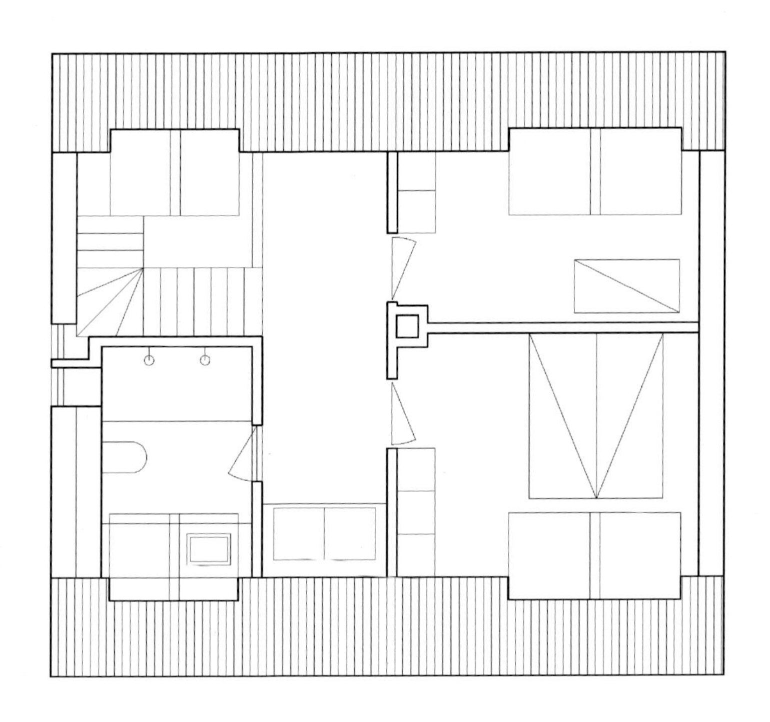

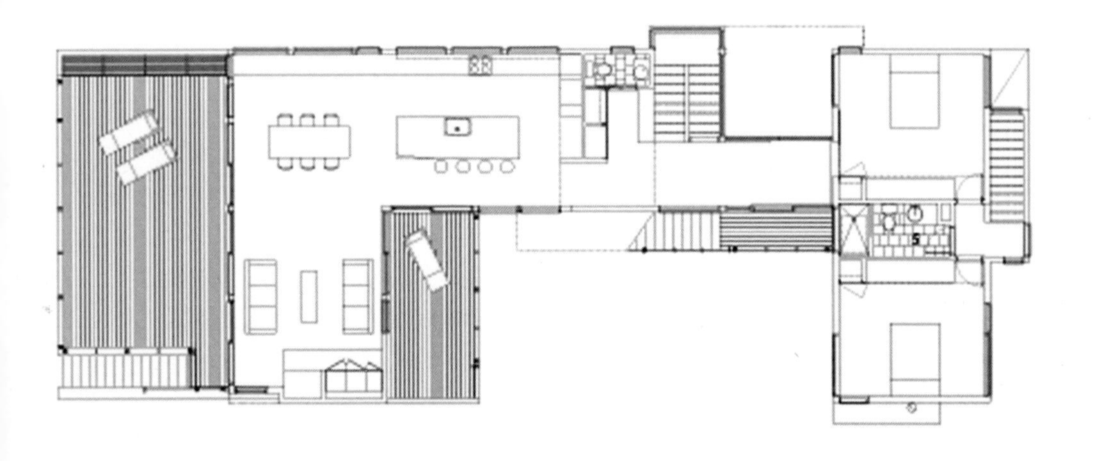

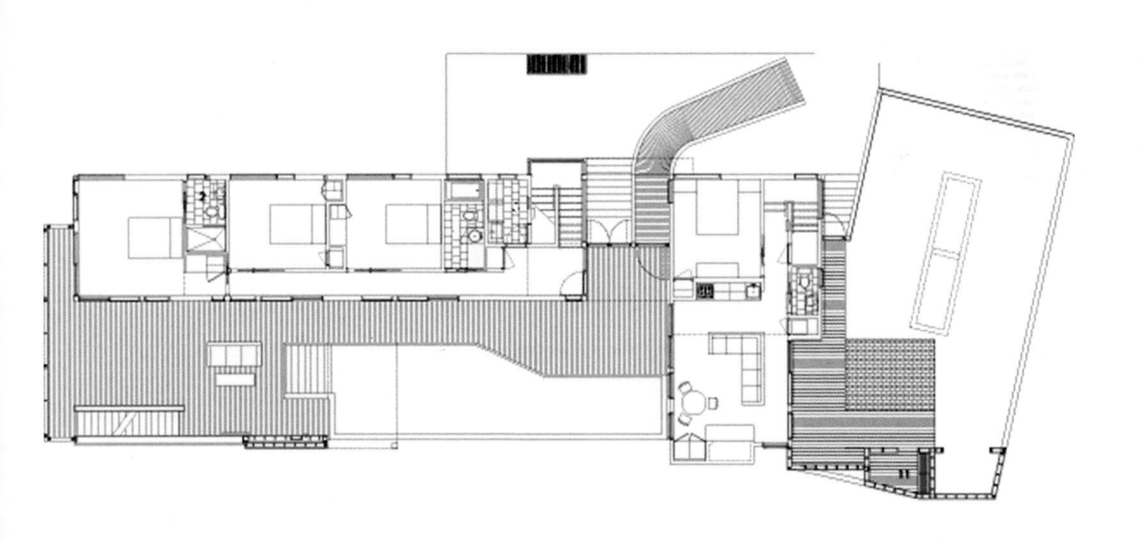

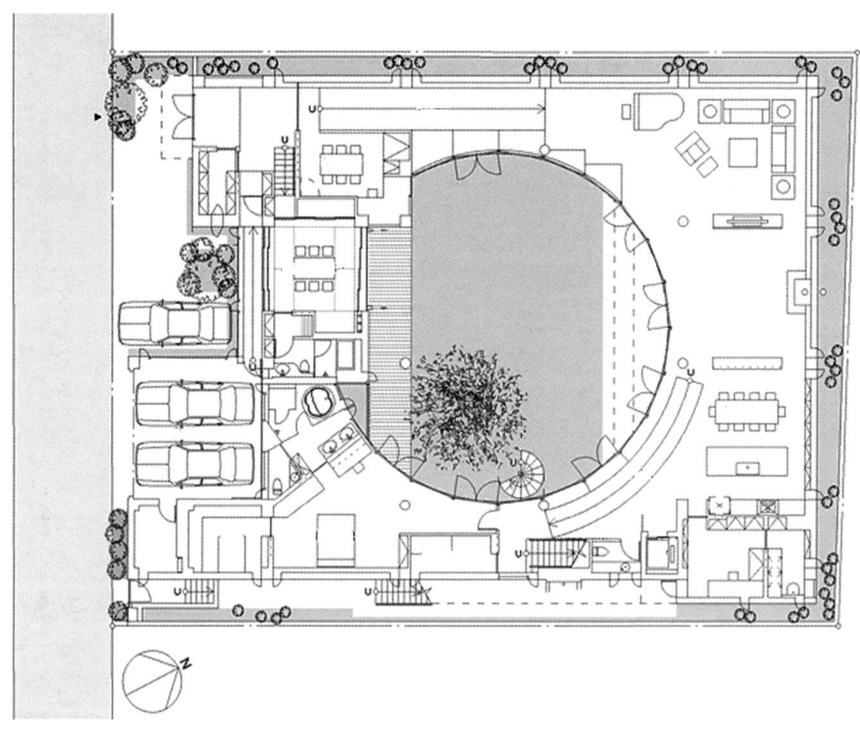

N

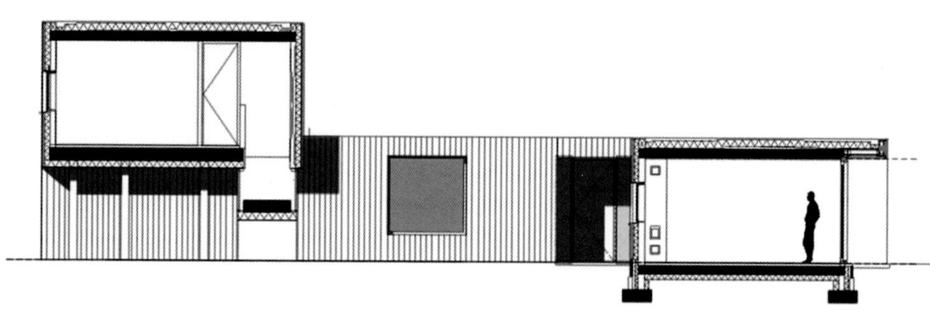

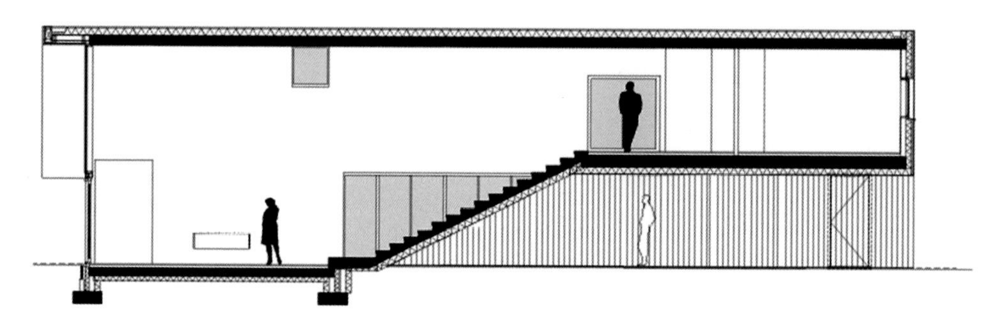

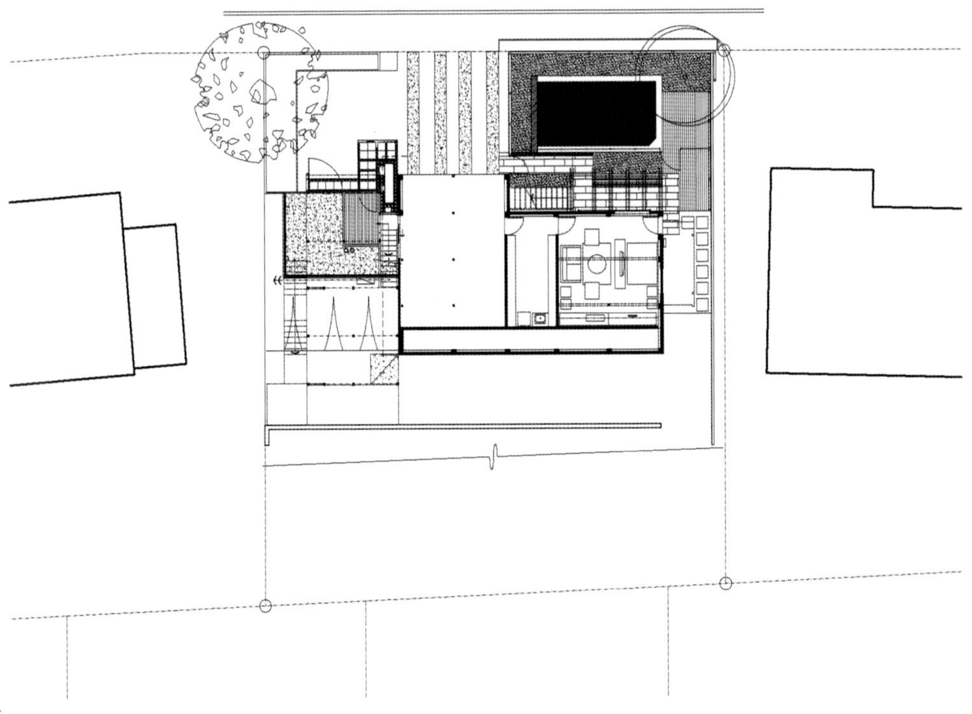

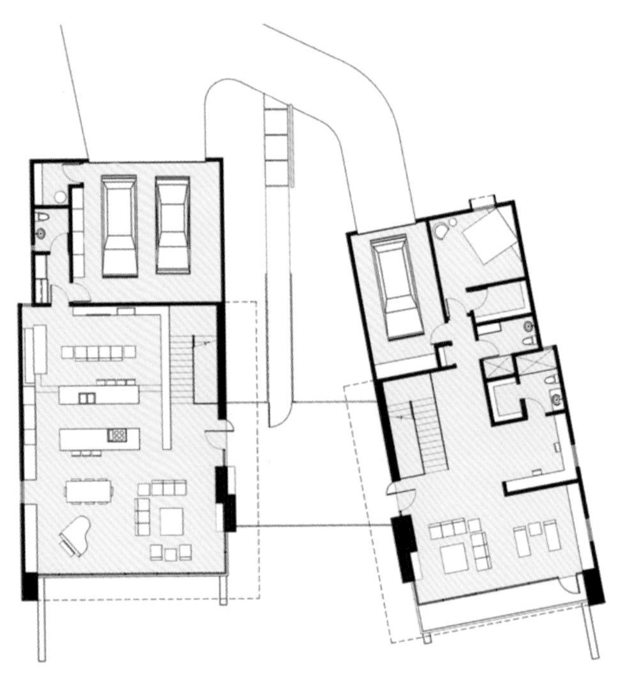

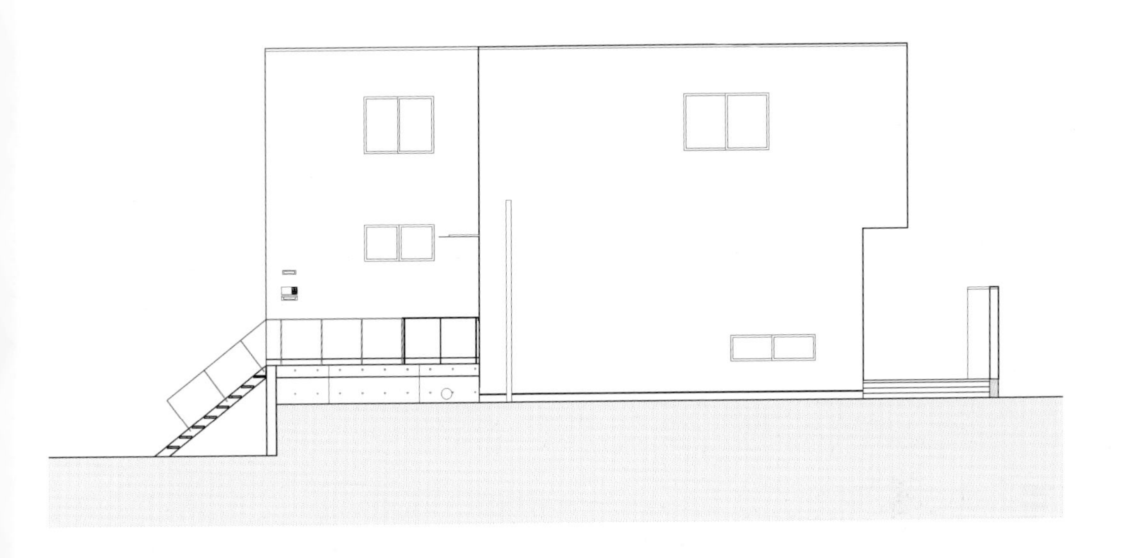

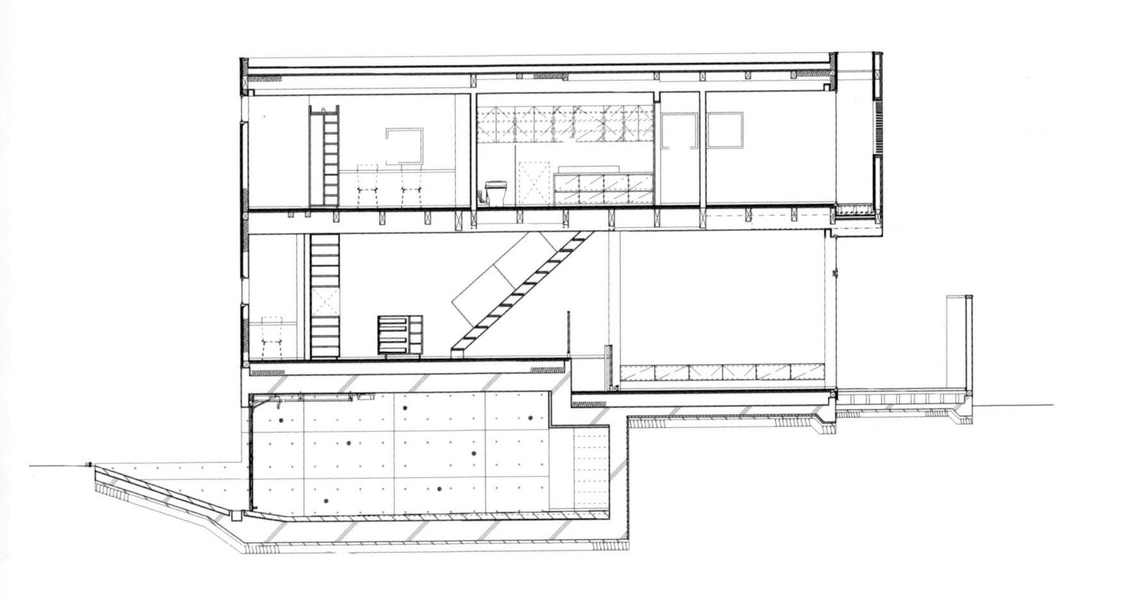

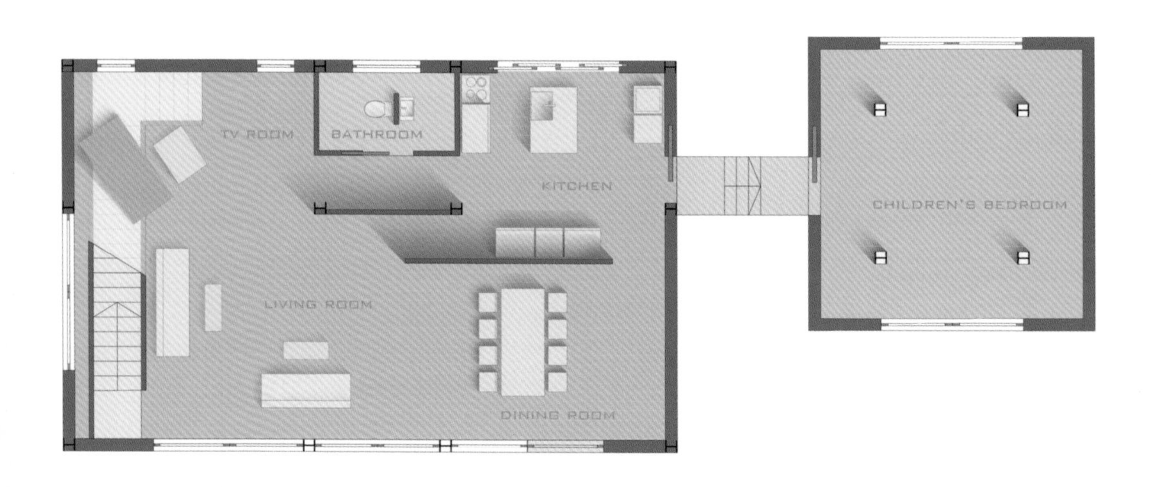
TV ROOM
BATHROOM
KITCHEN
CHILDREN'S BEDROOM
LIVING ROOM
DINING ROOM

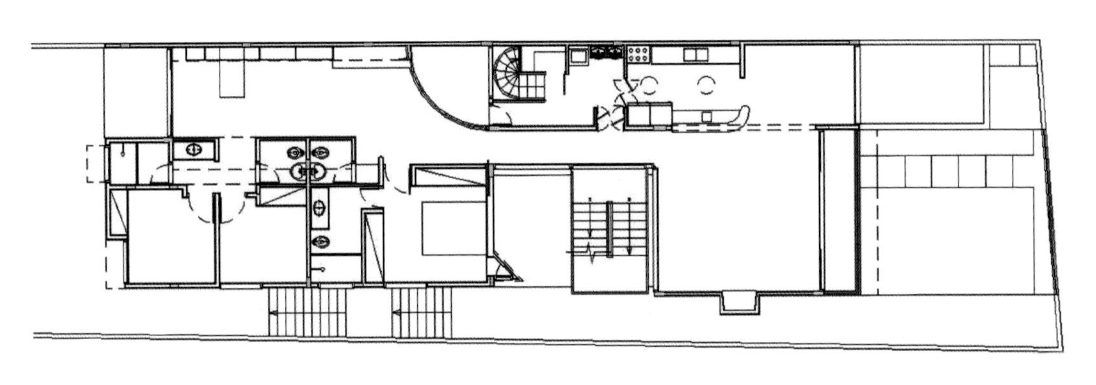

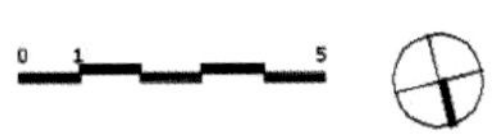
0
1
5

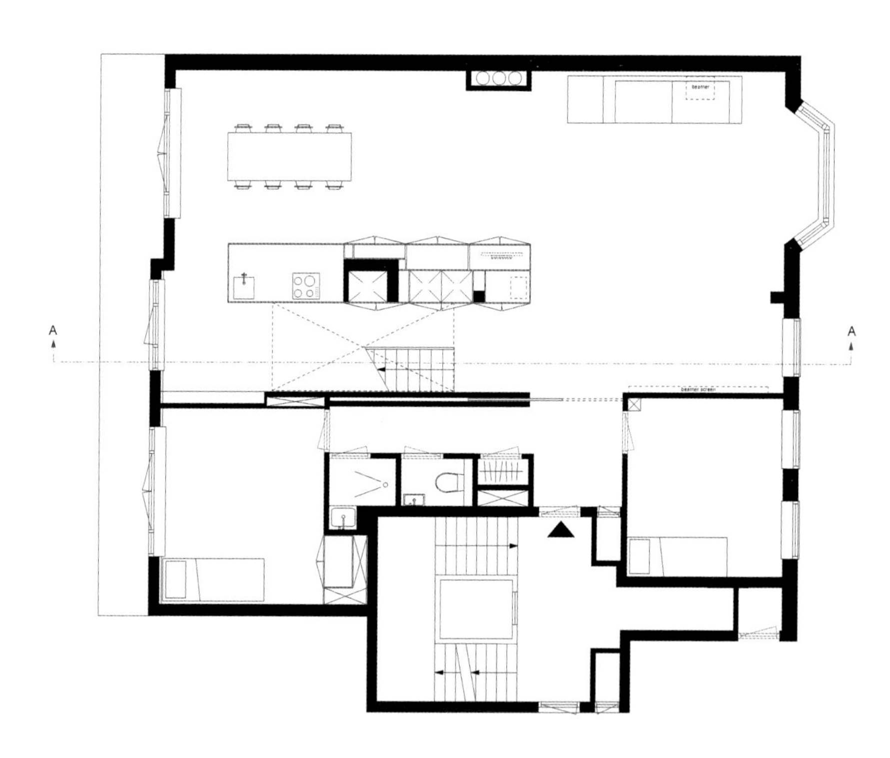
beamer
A
A
beamer screen

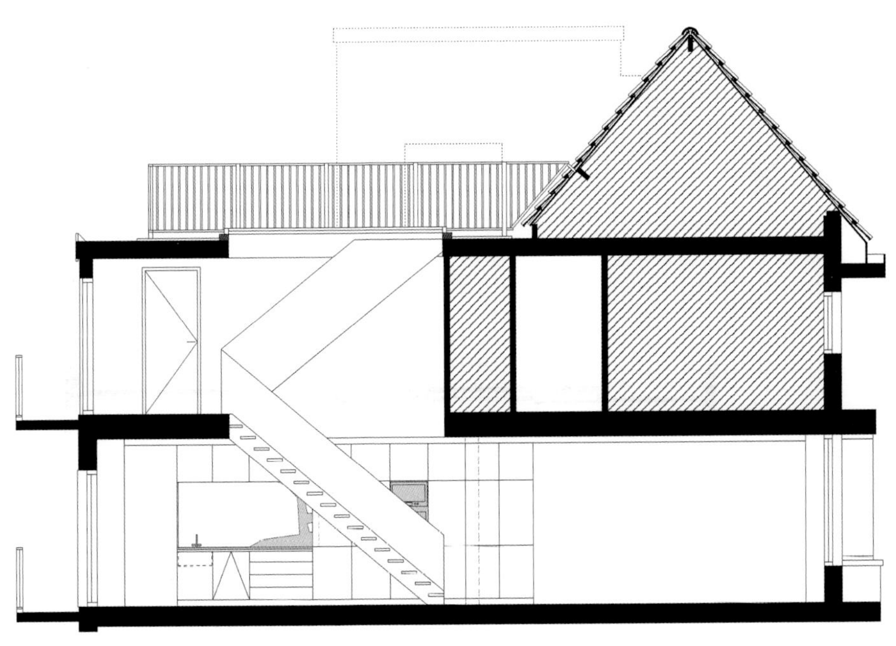

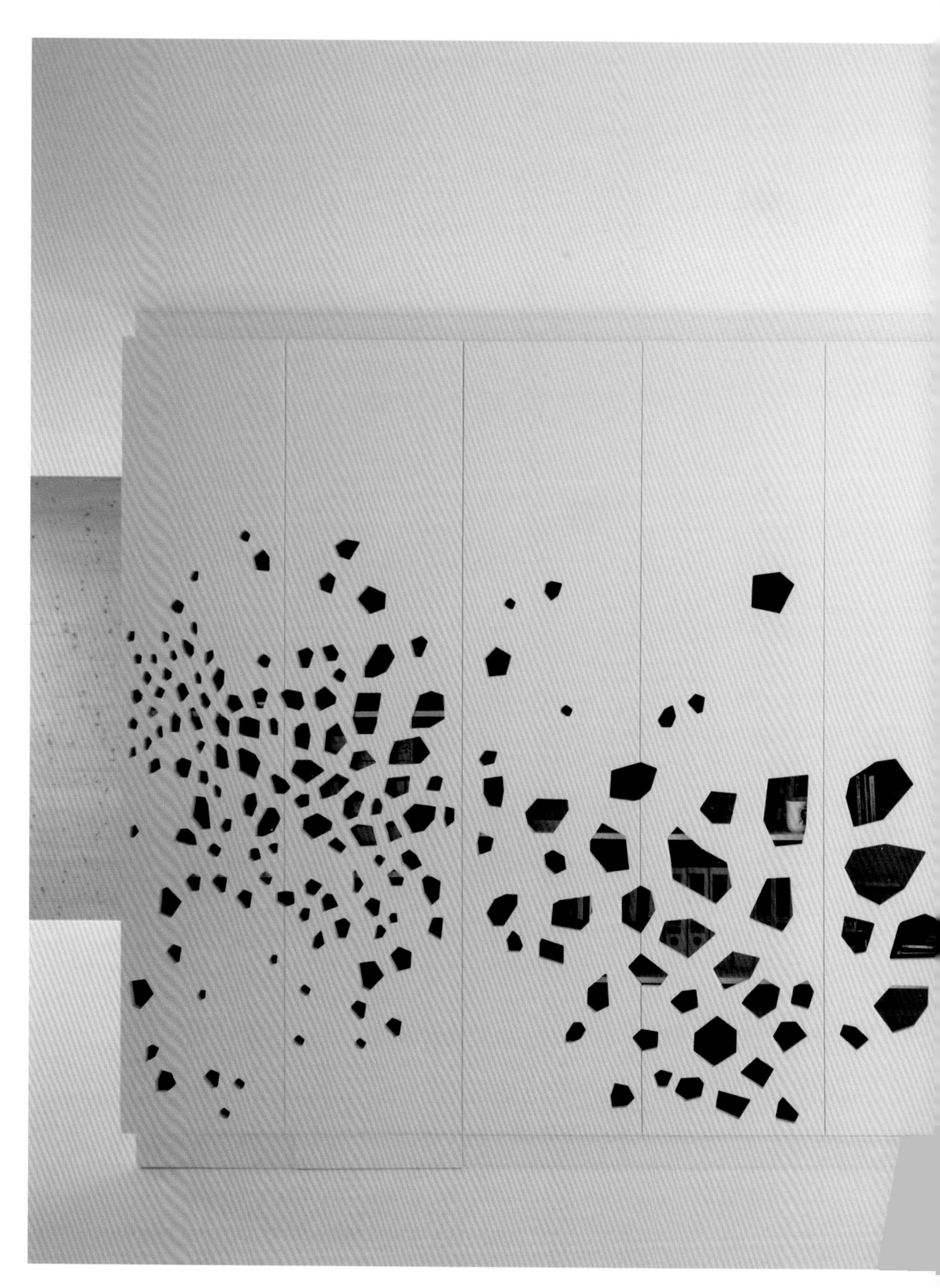

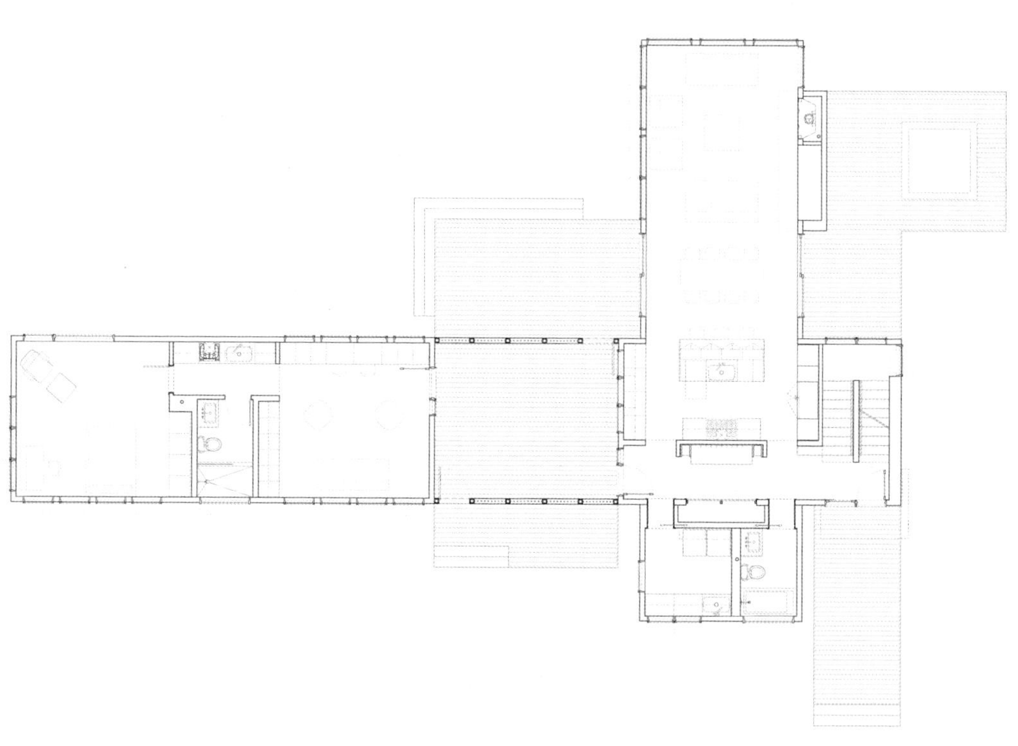

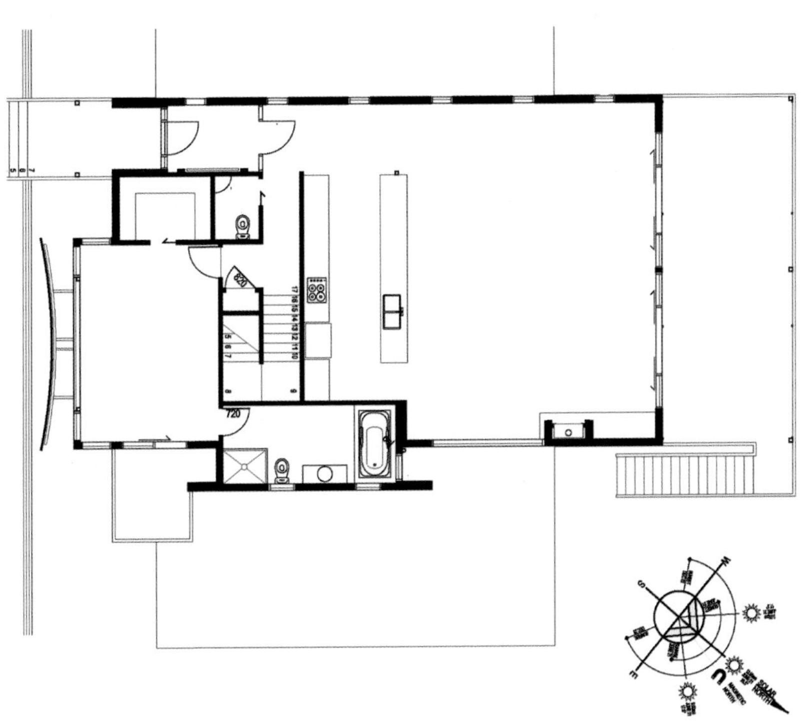

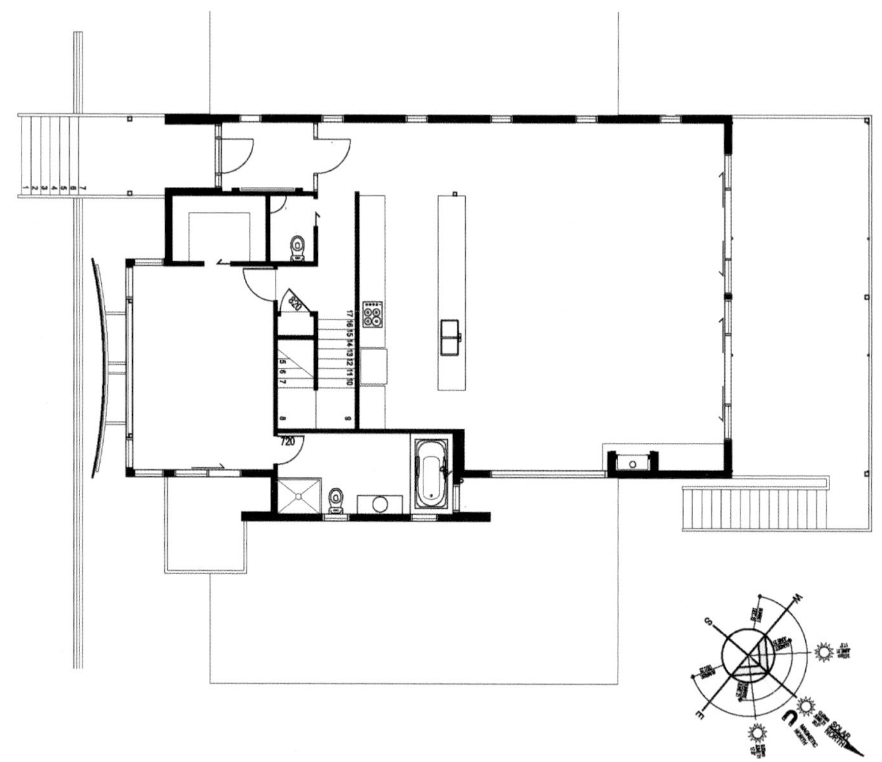

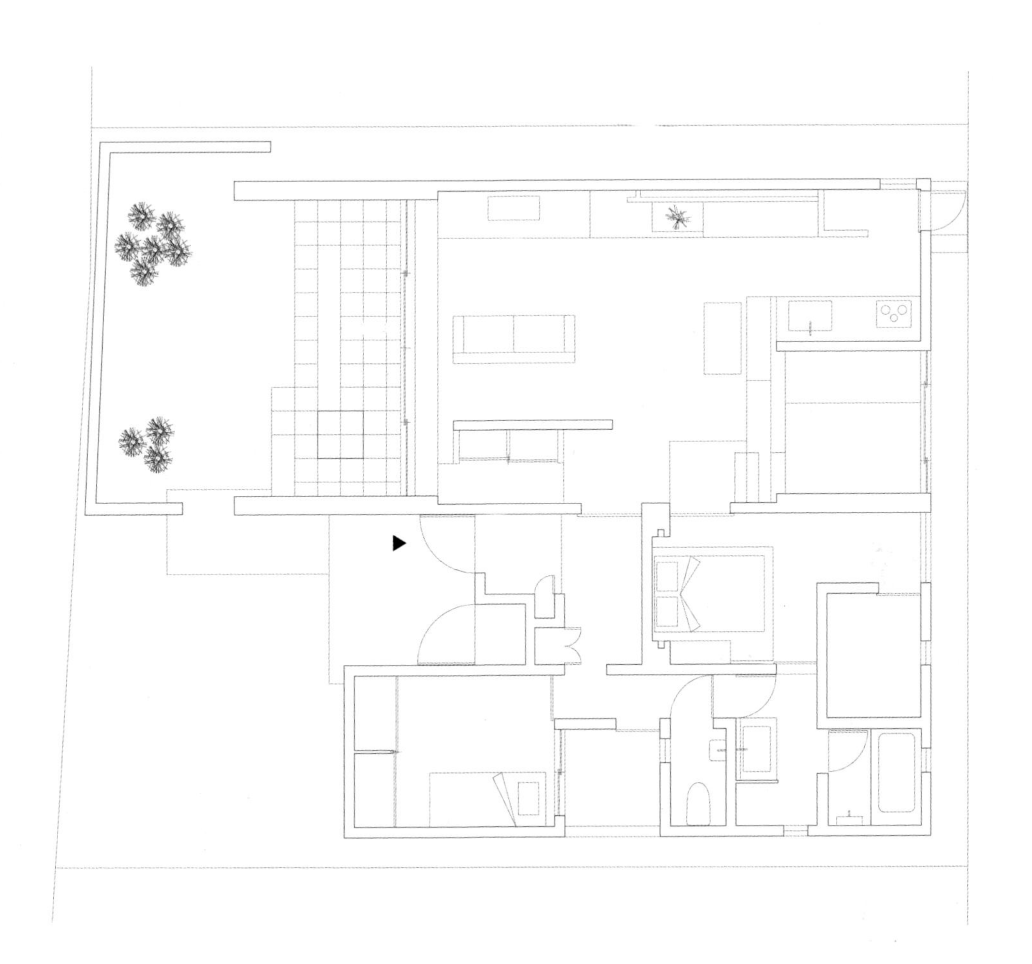

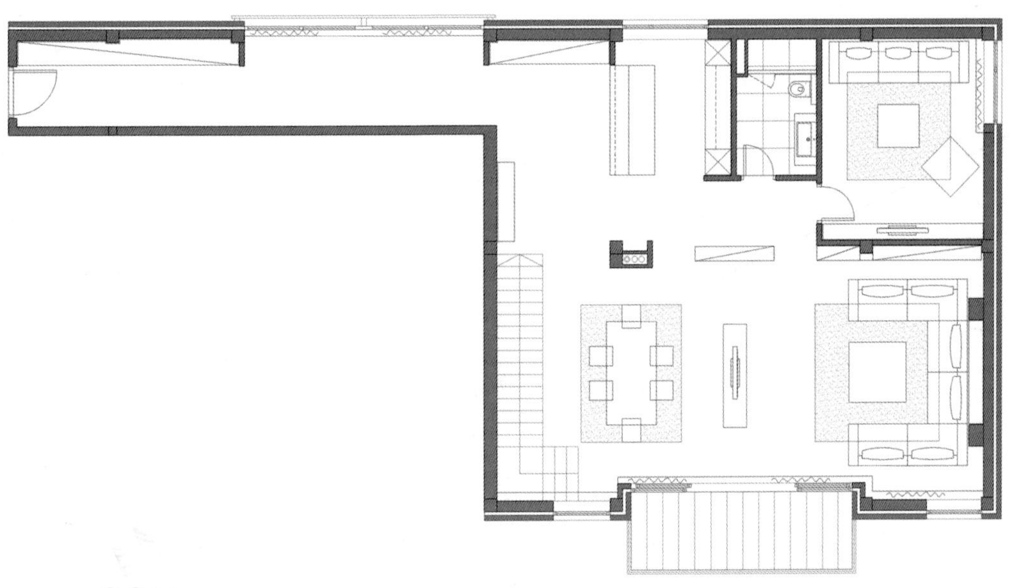

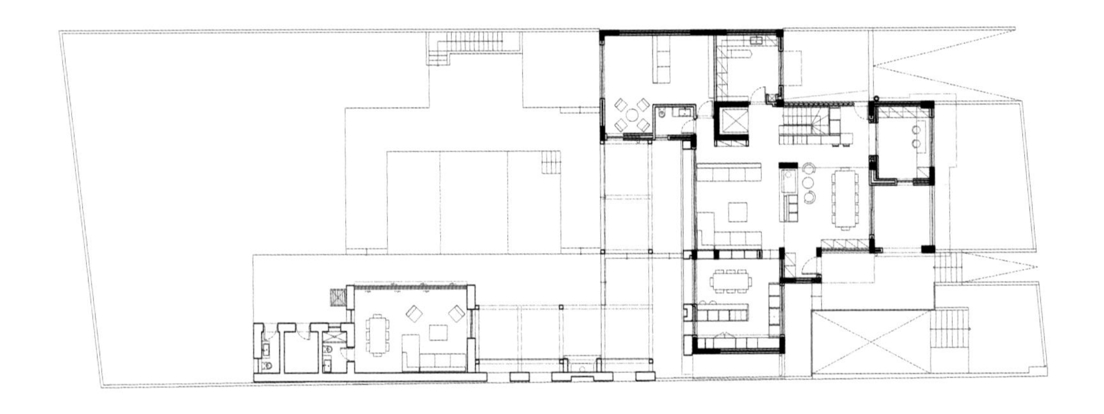

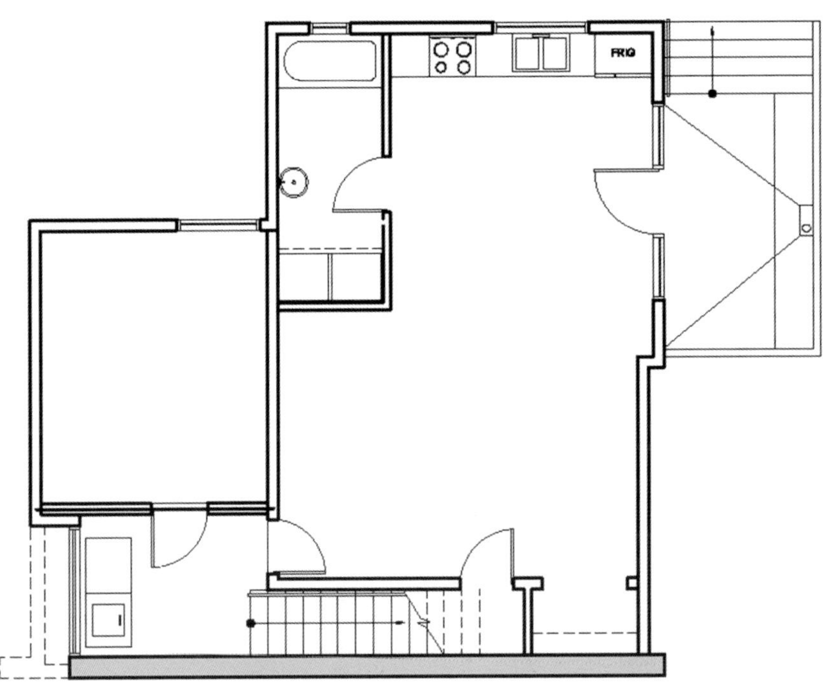
FRIG

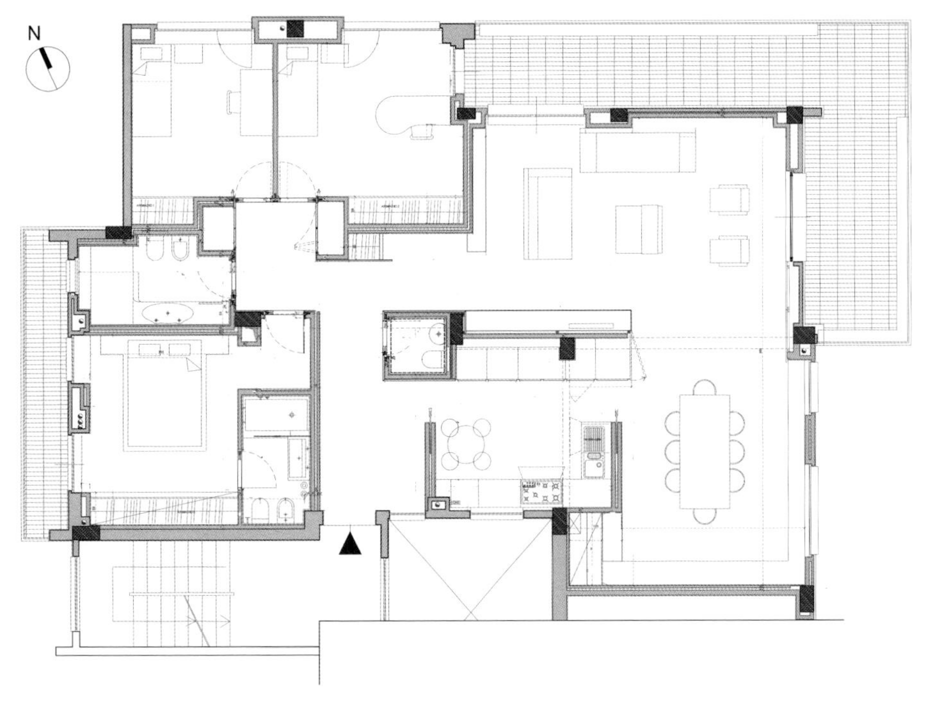
N

MARTINI

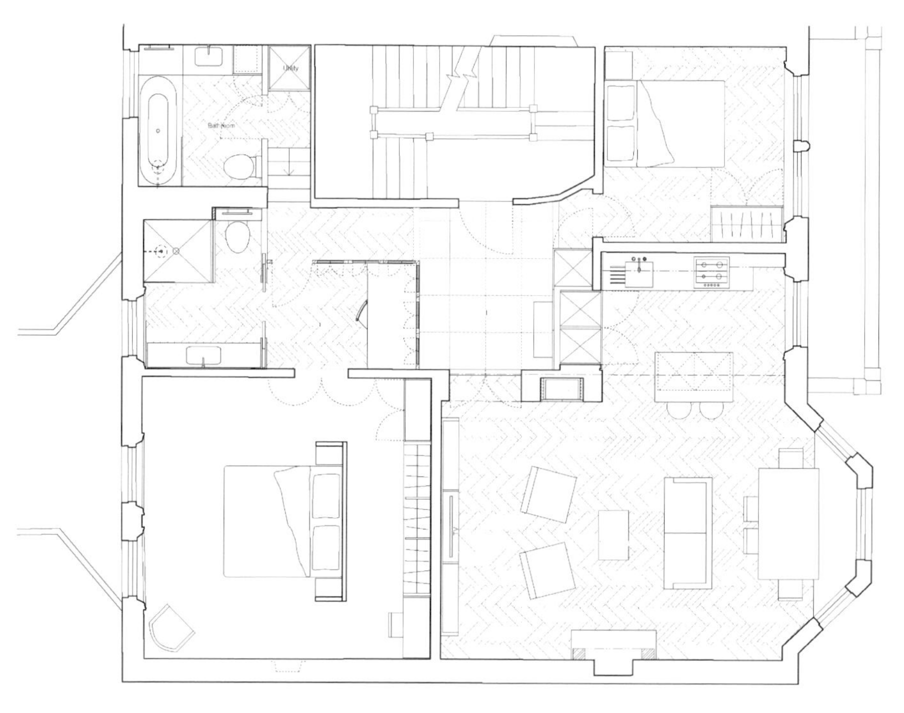
Bathroom

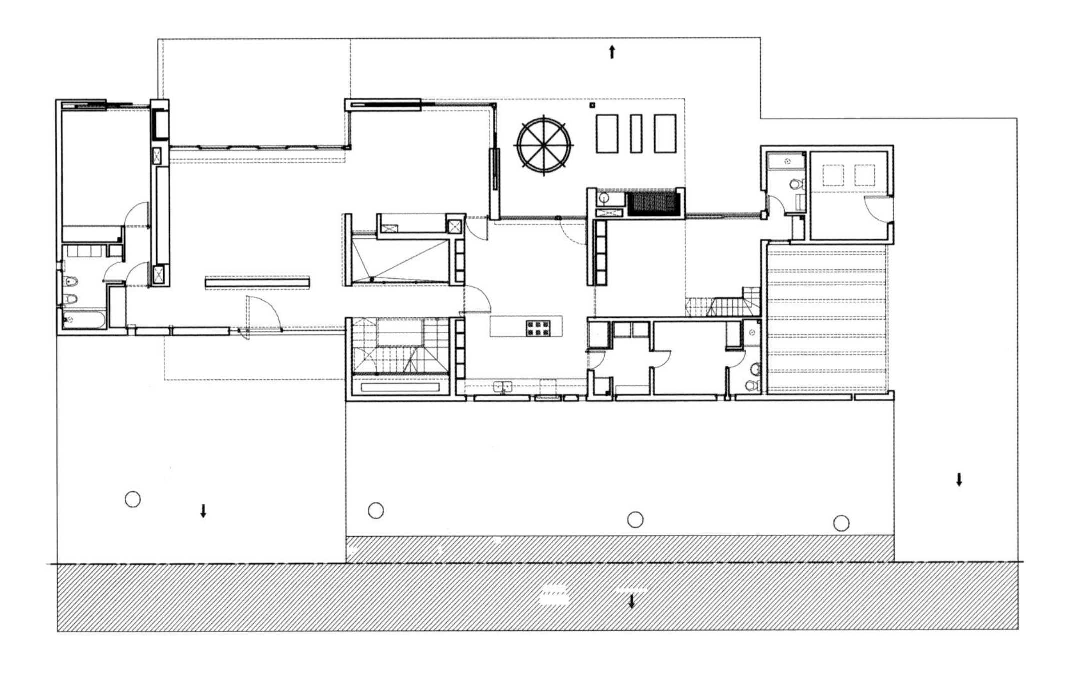

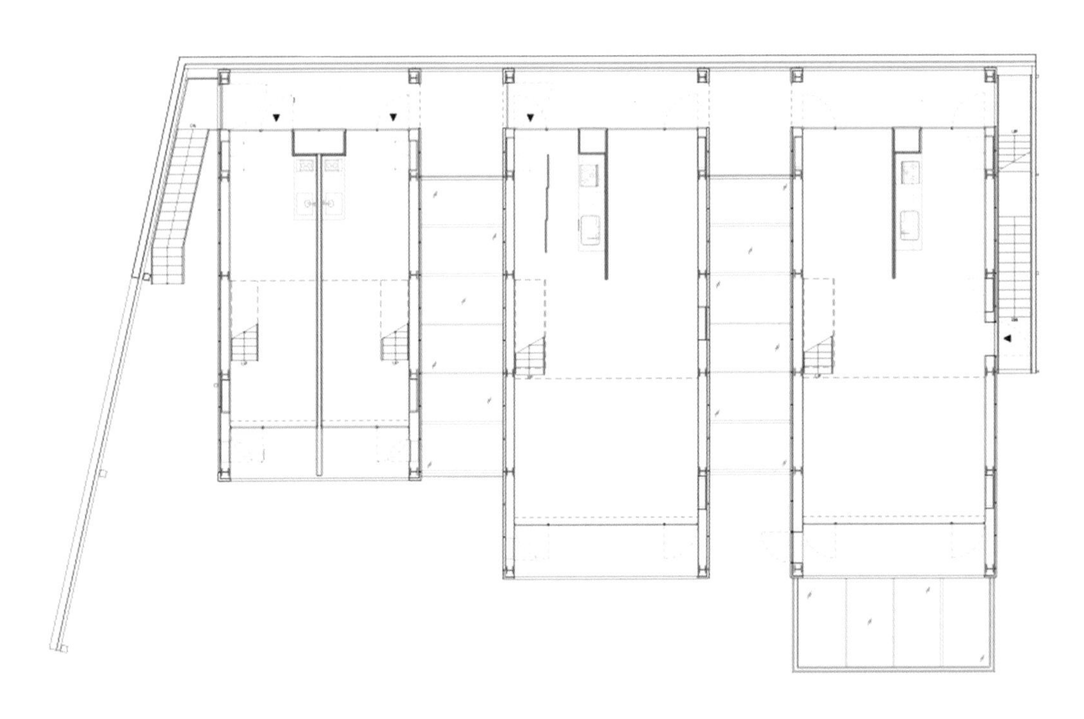

soup bible

PASTA

JENN-AIR

VIKING

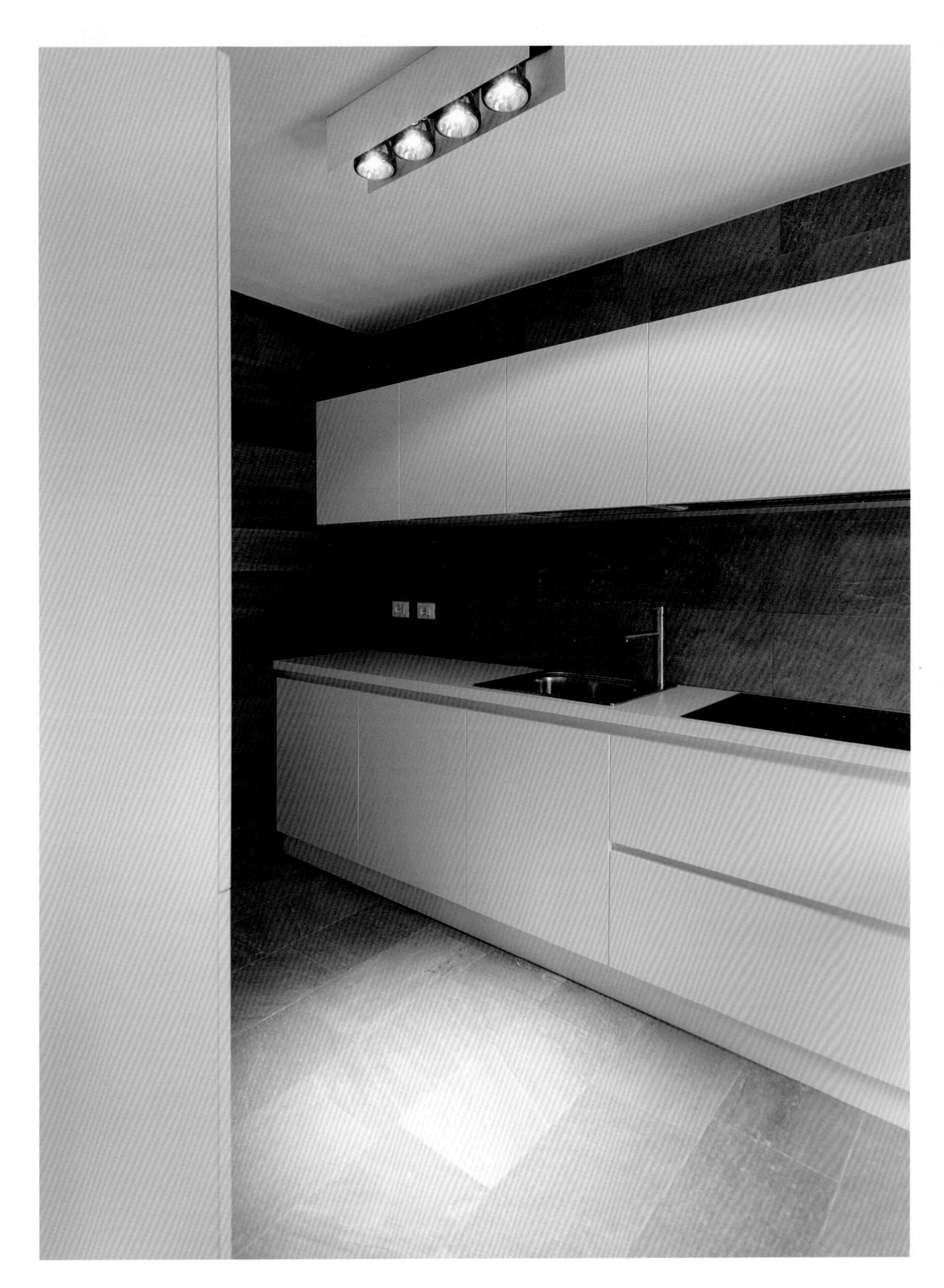

Index 索引

INDEX

+31ARCHITECTS,Amsterdam

3LHD

a21 studio

Affonso Risi Architect

Aleksandar Design Group

Aleksandar Savikin and Zoran Lazovic

Alex Choi

Altius Architecture Inc.

ALTUS Architecture + Design

Alvisi Kirimoto + Partners architecture

Andi Pepper Interior Design

ANDRES REMY ARQUITECTOS

ANDREW MAYNARD ARCHITECTS

Antonio Sofan

Arch. Giovanni Pellicciotta

Archikubik

ARCHITECT ELEMENTAL DESIGN, LLC

Architecture & Interior Design: Hariri & Hariri

Architecture: Graham Smith, Joe Knight

ArlicGalindez

ARTHUR CASAS

assemblageSTUDIO

Autoban

Base Architecture - Shawn Godwin

Belzberg Architects

BFLS - London Architects

Caroline Di Costa Architect and iredale pedersen hook architects

Christina Zerva Architects

Christopher Rose Architects

Coates Design Architects

Dane Richardson

Dane Richardson

David Collins Studio

Davor Mikulcic Dipl. Eng. Arch. (Sarajevo) RAIA, ANZIA

De Rosee Sa

Dennis Gibbens Architects

dep studio

Despang Architekten

Diane Pham

Edward Suzuki Associates (Edward Suzuki, Toshiharu Nanba)

Edwards Moore

Edy Hartono, Edha Architects

ELEMENTAL DESIGN, LLC

Ellivo Architects

Feldman Architecture

Fernanda Marques Arquitetos Associados

Ferrolan lab

Filippo Bombace

FORM/Kouichi Kimura Architects

Frédéric Haesevoets Architect

Friedrich St.Florian, FAIA

Gary Lee Partners

GIOVANNI VACCARINI architects

GLR arquitectos / Gilberto L. Rodríguez

GRAFT Gesellschaft von Architekten mbH, Berlin

Griffin Enright Architects

Grosfeld van der Velde Architecten

Hagy Belzberg

Hariri & Hariri-Architecture

Hofman Dujardin Architects

Hollwichkushner Llc (Whkn)

i29

II BY IV Design

Interface Studio Architects

IROJE KHM Architects

JANUS/MAC

Jean de Lessard

Joan Casals Pañella / CSLS arquitectes

Index 索引

Joey Ho / Joey Ho Design Ltd

John Friedman Alice Kimm Architects

Johnfriedmanalicekimmarchitects

Johnson Chou

Jonathan Marvel, Rob Rogers, Scott Demel and Nebil Gokcebay

Joseph Tanney, Robert Luntz

Juan Carlos Doblado

Judd Lysenko Marshall Architects

Katariina Rautiala, architect

Kathleen Hay Designs

Kathryn Scott

Kavellaris Urban Design

Kinari design

LEVEL Architects

Lightarchitecture

Louis Kahn Architect

LOVE architecture and urbanism

Luca Scacchetti

M. Ito Design, Inc

Mark Lind

Marmol Radziner

Martyn Lawrence Bullard

Max Strang Architecture

McKinney York Architects

MIGUEL ANGEL BORRAS & ELODIE GRAMMONT

Miro Rivera Architects

Monica Suarez & Christian von Düring

Monika Kozlowska

Montagna Lunga

Morrison Seifert Murphy

Naturehumaine

Ndarchitecture

Norm Architects / Jonas Bjerre-Poulsen & Kasper Rønn

Original Vision